COMPETITIVE BIOLOGY 1

INTRODUCTION

This objective biology series provides a basic and challenging problem of biology from particular topics. It can be used to brush up ones basics and checking up the preparation level of particular topic. It is equally helpful to the traditional classes as well as competitions. It can be also taken as a revision material for any competition which includes the test of basic biology. If you want to grasp the subject before practicing these multiple choice questions, you can go through the website http://www.ncert.nic.in/ncerts/textbook/textbook.htm and down load the free copy of science books and after having command on the topic practice it. For revision purpose, important points are given at the starting of each topic.

If you have any query or suggestion about this series you can send your suggestion at uk2594@gmail.com.

CONTENTS

CHAPTER 1   CELL

CHAPTER 2   TISSUE

CHAPTER 3   DIVERSITY IN LIVING ORGANISMS

CHAPTER 4   WHY DO WE FALL ILL

1. CELL

SOME IMPORTANT POINTS

- ➢ Robert hooked discover the cell in 1665 with the help of a primitive microscope.
- ➢ Cell is the fundamental unit of life.
- ➢ The plasma membrane allows the entry or exit of some material in and out from the cell.
- ➢ The movement of substances from a region of high concentration to the region low concentration is known as differs.
- ➢ The movement of water from a region of high concentration to the low concentration is known as osmosis.
- ➢ Plasma membrane is flexible and is made up of organic molecules called liquids and proteins.
- ➢ Amoeba gets its food from the external environment by the process of endnocytosis.
- ➢ Cellulose is a complex substance and provides structural strength to plants.
- ➢ Chromosomes contain information for inheritance of features from parents to next generation in the form of DNA (Deoxyribo Nucleic Acid).
- ➢ Mitochondria is known as the power house of cell. Because the energy needed for some activities in stored in it is the form of ATP.
- ➢ Plants cells have cell wall while animal cells have not cell wall.
- ➢ Lysosomes are known as the suicide bags of cell.
- ➢ Plastids are only present in plant cell.
- ➢ Plant cells have big vacuoles in compression to animal cell.
- ➢ The primary function of leucoplasts is storage.

1. CELL

1. The largest cell in the human body is?

   a) Nerve cell   b) Muscle cell

   c) Liver cell   d) Kidney cell

2. The barrier between the protoplasm and the other environment in an animal cell is?

   a) Cell wall   b) Nuclear membrane

   c) Neuron   d) Plasma membrane

3. A plant cell differs from an animal cell in the absence of?

   a) Endoplasmic   b) Mitochondria

   c) Ribosome   d) Centrioles

4. Centrosome is found in?

   a) Nucleus   b) Cytoplasm

   c) Chromosomes   d) Nucleus

5. The power house of a cell is?

   a) Chloroplast   b) Golgi apparatus

   c) Nucleus   d) Mitochondria

6. Within a cell the site of respiration is the?

   a) Ribosome   b) Golgi apparatus

   c) Endoplasmic reticulum   d) Mitochondria

7. Which is called suicidal bag of cell?

   a) Centrosome   b) Lysosome

   c) Mesosome   d) Chromosome

8. Ribosome is the centre for?

a) Protien synthesis   b) Respiration

c) Fat synthesis   d) None of these

9.   Double membrane is absent in?

a) Mitochondria   b) Chloroplast

c) Lysosome   d) Nucleus

10.   Cell organelle found only in plant is?

a) Golgi apparatus   b) Mitochondria

c) Plastids   d) Ribosome

11.   Organisms lacking nucleus and membrane bound organelle are?

a) Diploids   b) Haploids

c) Eukaryotes   d) Prokaryotes

12.   Animal cell is limited by?

a) Basement membrane   b) Plasma Membrane

c) Cell Wall   d) Shell Membrane

13.   The Network of Endoplasmic Reticulum is present in the?

a) Nucleus   b) Nucleolus   c) cytoplasm   d) Chromosomes

14.   Lysosomes are reservoirs of?

a) Fat   b) RNA

c) Secretary Glycoprotein   d) Hydrolytic Enzymes

15.   The membrane surrounding the vacuole of a plant cell is called?

a) Tonoplast   b) Plasma Membrane

c) Nuclear Membrane   d) cell wall

16.   Cell Secretion is done by?

a) Plastids   b) ER   c) Golgi   d) Nucleolus

17. Centrioles are associated with?

a) DNA synthesic   b) Reproduction

c) Spindle formation   d) Respiration

18. Main difference between animal cell and plant cell is due to?

a) chromosome   b) Ribosome

c) Lysosome   d) none of these

19. Animal cell lacking nuclei would also lacking?

a) Chromosome   b) Ribosome   c) Lysosome   d) none of these

20. Plasmolysis occurs due to?

a) Absorption   b) Endosmosis

c) osmosis   d) Exosmosis

21. A plant cell becomes turgid due to?

a) Plasmolysis   b) Exdosmasis

c) Endosmosis   d) Electrolysis

22. Solute Concentration is higher in the external solution called?

a) Hypotonic   b) Isotonic

c) Hypertonic   d) None of these

23. A cell placed in Hypertonic solution will?

a) Swell up   b) Shrink

c) No Change in shape, size   d) show plasmolysis

24. The radiant energy of sunlight is converted to chemical energy and is stored as?

a) AMP   b) ADP

c) ATP   d) APP

25. Which of the following organelle does not have membrane?

a) Ribosome

b) Nucleus

c) Chloroplast

d) none of these

26. Root hair absorbs water from soil through?

a) Osmosis

b) Active transport

c) Diffusion

d) Endocytosis

27. The number of lenses in compound light microscope is?

a) 4   b) 3   c) 1   d) 2

28. Which cell organelle is not bounded by a membrane?

a) Ribosome   b) Lysosome   c) ER   d) Nucleus

29. Which of the following cellular part possess a double membrane?

a) Nucleus   b) Chloroplast   c) Mitochondrion d) All of the above

30. Cristae and Oxysomes are associated with?

a) Mitochondria

b) Plastids

c) Plasma Membrane

d) Golgi Apparatus

31. Karyotheca is another name of?

a) Nuclear envelope   b) Nucleus c) Nuclear pores d) Nucleolus

32. Cell organelle that acts as supporting skeletal framework of the cell is?

a) Golgi apparatus   b) Nucleus c) Mitochondria   d) ER

33. Plastids are present in?

a) Animal cell only

b) Plant cell only

c) Both

d) None of these

34. Cell wall of plant is chiefly composed of?

a) Hemi cellulose   b) Cellulose c) Proteins   d) none of these

35. Intercellular connections of plant cells are called

a) Middle lamella   b) Micro fibrils   c) Matrix   d) Plasmodesmata

36. Genes are located on the?

a) Chromosomes   b) Nucleoulus

c) Nuclear membrane   d) Plasma membrane

37. Chromatin Consists of?

a) RNA   b) DNA   c) RNA& Histones   d) DNA & Histones

38. Name of the process that requires energy provided by ATP?

a) Diffusion   b) Osmosis   c) Active Transport   d) Plasmolysis

39. Vacuoles are present in?

a) Animal cells   b) Plant Cells   c) Both   d) None of these

40. The cell is the fundamental structural unit of?

a) Non living organisms   b) Living organisms

c) Both   d) None of these

41. Who saw that the cork resembled the structure of a honeycomb consisting of many little compartments?

a) Robert Hooke   b) N.S. Bose c) M.K. Chandran   d) None of these

42. ------------ is a substance which comes from the bark of a tree?

a) Cork   b) torck   c) Metal   d) None of these

43. Cell is a latin word for?

a) a little room   b) a big room

c) a little basement   d) None of these

44. Hook made ----------- chance observation through a self designed microscope?

a) Slice of cork   b) Small of cork   c) None of these

45. The very first time that someone had observed that living things apper to consist of separate units?

a) Hooke   b) Einstein   c) Golgi   d) none of these

46. Cell was first discovered by Robert Hooke in?

a) 1666   b) 1665   c) 1667   d) 1668

47. Who discovered the nucleus in the cell?

a) Robert Hook   b) Leeuwen Hook   c) Robert Brown   d) None of these

48. Purkinje in ----------- coined the term protoplasm for the fluid substance of the cell?

a) 1837   b) 1836   c) 1839   d) None of these

49. Which theory tells that all the plants and animals are composed of cells and that the cell is the basic unit of life?

a) Cell theory   b) Plant theory

c) Animal theory   d) None of these

50. Group of cells having similar structure and function are termed as?

a) Tissue   b) Organ System   c) Organ   d) Living organism

Answers:

QUES.	ANS.	QUES.	ANS.	QUES.	ANS.	QUES.	ANS.	QUES.	ANS.
1. (a)	2.(d)	3. (d)	4. (b)	5. (d)	6. (d)	7. (b)	8. (a)	9. (c)	10. (c)
11. (d)	12. (b)	13.(c)	14. (d)	15. (a)	16. (c)	17. (c)	18. (a)	19. (a)	20. (d)
21. (c)	22. (c)	23. (a)	24. (c)	25. (a)	26. (a)	27. (d)	28. (a)	29. (d)	30. (a)
31. (a)	32. (d)	33. (b)	34. (a)	35. (d)	36. (a)	37. (d)	38. (c)	39 (b)	40. (b)
41. (a)	42. (a)	43. (a)	44. (a)	45. (a)	46. (b)	47. (c)	48. (c)	49. (a)	50. (a)

2. TISSUE

SOME IMPORTANT POINTS

➤ Tissues are made up of cells.
➤ Millions of cells carried out a particular function in the body hence this group of cells is known as tissue.
➤ Mainly plant tissues are of two types :
1. Meristmetic tissue
2. Permanent tissue
➤ Merismetic tissue is divided into three parts.
1. Apical meristem
2. Inter calary meristem
3. Lateral meristem
➤ Apical meristem is found at the growing tips of roots and shoot to increase the length of stem and root.
➤ Permanent tissue are divided into two types :
1. Simple permanent tissue
2. Complex permanent tissue
➤ Parenchyma, Collenchyma, choleranchyma and sclerenchyma are the types of simple permanent tissue.
➤ Xylem and phloem are the type of complex permanent tissue.
➤ Xylem transports water and minerals from the root to the whole body and phloem transports food from leaves to the whole body of the plant.
➤ Animal tissues are mainly of four types :
1. Epithelial tissue
2. Connective tissue
3. Muscular tissue
4. Nervous tissue
➤ According to the structure epithelial tissue are divided as squamous, cuboidal, columnar, ciliated, and glandular.

- Different types of connective tissue are found in our body includes areolar tissue, adipose tissue, bone, tendon, cartilage and blood.
- Nervous is the functional unit of nervous tissue.
- Nervous tissue receives and conducts impulse. The impulses are read by brain.

2. Tissue

1.   The unicellular organism has?

(a)Single cell (b) Multiple cell   (c) Both (a) & (b)   (d) None of these

2.   The multicellular organism is mainly?

(a)Man   (b) Yeast   (c) Bacteria   (d) None of these

3.   The activities perform by single called organism are?

(a)Digestion   (b) Movements   (c) Both (a) & (b)   (d) None of these

4.   A tissue may be defined as a?

(a)Group of cells (b)Single cells (c)Both(a)&(b)   (d)None of these

5.   The cell walls of meristematic tissue are made up of?

(a)Cellulose   (b) Protien   (c)Fats   (d) None of these

6.   The lateral meristem is located at the growing?

(a)Apices   (b)Leaf   (c)Stem   (d) None of these

7.   The merestimatic tissue are responsible for?

(a)Linear growth of an organ   (b) Increase in diameter

 (c)Growth of leaves   (d) None of these

8.   The intercalary meristem responsible for?

(a)Growth of leaves (b)Growth of leaves   (c)Both(a)&(b)   (d) None of these

9.   The living permanent cells are large and posses?

(a)Thin walls   (b)Thick walls   (c)Both(a)&(b)   (d) None of these

10.   Permanent tissue may be at?

(a)Fixed position   (b)Dividing continuously   (c)Both (a)&(b) (d) None of these

11.   Food and water are conducted in plants through?

(a)Vascular tissue (b)Mascular tissue (c)Both(a)&(b) (d) None of these

12.   The supported tissue are mainly present in?

(a)Animals   (b)Plants   (C)Both(a)&(b) (d) None of these

13.   Differentiation in the process of taking up a permanent?

(a)Shape   (b)size   (c)Function   (d) None of these

14.   Simple permanent tissue are composed of?

(a)Only one type of cell (b)Variety of cell (c)Mixed cell (d) None of these

15.   The flexibility in plants is due to the?

(a)Collenchyma (b)Parenchyma (c)Sclerenchyma   (d) None of these

16.   In aquatic plants byouancy by which plants can float is provided by?

(a)parenchyma (b)Sclerenchyma (c)Collenchyma (d) None of these

17.   To perform photosynthesis chlorophyll is contain in?

(a)Chlorenchyma (b)Collenchyma (c)Sclernchyma (d) None of these

18.   Easy bending in different parts of plants without breaking is due to?

(a)Collenchyma (b)Parenchyma (c)Chlorenchyma (d) None of these

19.   Collenchyma provided some functions to the organs where it is found these

Functions are?

(a)Flexibility (b)Elasticity (c)Both(a)&(b) (d) None of these

20.   Sclerenchyma tissue help the plant to make it?

(a)Hard   (b)Soft   (c)Sniff   (d) None of these

21. The husk of a coconut is made up of?

(a) Parenchyma (b) Chlorenchyma   (c) Collenchyma (d)Sclerenchyma

22. The walls of sclerenchyma tissue are thickened due to:
(a)Proteins   (b)Lignin   (c)Chloroplast   (d)None of these.

23. Sclerenclhyma tissue is present in:

(a)Stems   (b)The covering of seeds,nuts

(c)Veins of leaves   (d) All of these

24. A little or no number of protoplasm is present in:

(a)Parenchyma tissue   (b)Scelerenchyma tissue

(c)Chlorenchyma tissue   (d) None of these

25. The outermost cell layer of the cell is ?

(a)Epidermis (b)Cytoplasm (c)Stomata (d) None of these

26. Cells of epidermis on the ariel parts helps again?

(a)Loss of water (b)Cytoplasm (c)Both(a)&(b) (d) None of these

27. Small pores in the epidermis of leaf are?

(a)Cytoplasm (b)Stomata (c)Guard cell   (d) None of these

28. Each stomata is guarded with two kidney shaped cell called?

(a)Epidermis cells (b)Plants cell (c)Guard cell (d) None of these

29. The opening and closing of stomata is regulated by?

(a)Epidermal cell   (b)Plant cell (c)Guard cell   (d) None of these

30.   The cell of cork of a tree are?

(a)Dead   (b)A live   (c)Small   (d) None of these

31.   As plants grow older epidermis of stems are replaced by?

(a)Secondary meristem (b)Cytoplasm (c)Cork (d) None of these

32.   The chemical that makes the cork wall impervious to gases and water is?

(a)Suberin   (b)Ammonium   (c)Both(a)&(b)   (d) None of these

33.   Complex tissue are composed of?

(a)One type of cell   (b) More than one type of cell

(c) Mixed type of cell   (d) Both(b)&(c)

34.   The minerals and water from root to different parts of plants is conducted by

(a)Xylem   (b)Pholem   (c)Branches   (d) None of these

35.   The food material from the leaves to different parts of plants is conduct by?

(a)Xylem   (b)Pholem   (c)Roots   (d) None of these

36.   The parts of xylem in a plant are?

(a)Tracheids (b)Vessels (c)Xylem ,Parenchyma (d)All of these

37.   The food stored in the form of starch fat in ?

(a)Xylem Parenchyma   (b) Vessels (c) Tracheids   (d) None of these

38.   The constitute of phloem are?

(a)Sieve tube (b)Companion   (c)Vessels   (d)Both(a)&(b)

39.   Xylem consist both living and non-living?

(a)Beings   (b)Cells   (c)Chloroplast   (d) None of these

40.   In phloem food material from the leaves carry to storage organ and then

growing region of plant body by?

(a)Sieve tube (b)Pollen tube (c)Conducting tube   (d) None of these

41.   The passage of materials is controlled by?

(a)Conducting cell (b)Companion cell   (c)Epidermal cell (d) None of these

42.   The tissue found in our body ?

(a)Blood   (b)Xylem   (c)Muscles (d) Both(a)&(b)

43.   Blood is a which type of tissue?

(a)Epithelial tissue (b)Connective tissue (c)Muscular tissue (d) None of these

44.   The covering or protecting tissue in the animal body are?

(a)Epithelial tissue   (b)Connective tissue (c)Nervous tissue (d) None of these

45.   The skin which protects the body is made up of?

(a)Squamous epethilial   (b)Cuboidal epithelial

(c)Columnar epethilial   (d) None of these

46.   In the respiratory tract the tissue that have cilia with pair like projection is?

(a)Squamous epithelial   (b)Stratified squamous

(c)Columnar epithelial   (d) None of these

47.   The tissue that makes the lings of kidney tubules ,ducts,of salivary glands is?

(a)squamous epithelium   (b)Cuboidal tissue

(c)Columnar epithelium   (d)) None of these

48.   Bone a hard and non flexible tissue is an example of?

(a)Columnar tissue   b)Connecting tissue

(c)Nervous tissue   (d)) None of these

49. Blood has been a fluid matrix called ?

    (a)Plsma   (b)RBC   (c)WBC   (d)Protiens

50. Two bones can be connected to each other by connective tissue called?

    (a)Ligament   (b)Tendons   (c)Muscles   (d)) None of these

51. The tissue connects muscles to bones ?

    (a)Ligament   (b)Tendon   (c)Muscles   (d)) None of these

52. The tough ,smooth and flexible connective tissue hat present in nose ,ears
    trachea is ?

    (a)Ligament   (b)Cartridge   (c)Tendons   (d)) None of these

53. The tissue that found between the skin and muscles in the bone marrow ,
    Supports internal organs is ?

    (a)Ligament   (b)Cartridge   (c)Areolar   (d)) None of these

54. The tissue that found below the skin and store fats of our body is?

    (a)Ligament   (b)Adipose   (c)Tendons   (d)) None of these

55. Which tissue cause movement in our body?

    (a)Supporting tissue   (b)Mascular tissue

     (c)Nervous tissue   (d)) None of these

56. Example of voluntary action muscle is?

    (a)Muscles in our body   (b)Contratiction of blood vessel

    (c)Sneezing   (d)) None of these

57. The example of involuntary action muscle is?

    (a)Muscles in our limbs (b)The iris of eye (c)Lymph (d)Both (a)&(b)

58. The involuntary muscles of heart is called?

    (a)Lymph   (b)Cardiac

    (c)Neuron muscles   (d)) None of these

59. The compositons of tissue are?

    (a)Brain   (b)Spinal cord   (c)Nerves   (d)All of these

60. The cells of Nervous tissue is called?

    (a)Neurons   (b)Nerve cells

     (c)Both (a)&(b)   (d)) None of these

61. The longest cell of our body is?

    (a)Neuron   (b)Lymph

    (c)Forelimp   (d)) None of these

62. A Nerve cell consists of different parts?

    (a)Cell body   (b)Axon

    (c)Dendrites   (d)All of these

Answers :

Q	A	Q	A	Q	A	Q	A	Q	A	Q	A	Q	A	Q	A	Q	A	Q	A
1	A	8	A	15	A	22	B	29	C	36	D	43	B	50	A	57	D
2	A	9	A	16	A	23	D	30	A	37	A	44	A	51	B	58	B
3	A	10	A	17	A	24	B	31	A	38	D	45	A	52	B	59	D
4	A	11	A	18	A	25	A	32	A	39	B	46	C	53	C	60	C
5	A	12	B	19	C	26	C	33	B	40	A	47	B	54	B	61	A
6	C	13	D	20	D	27	B	34	A	41	B	48	B	55	B	62	D
7	A	14	A	21	D	28	B	35	B	42	D	49	A	56	A		

3. DIVERSITY IN LVING ORGANISMS

SOME IMPORTANT POINTS

➢ Classification helps us to read the diversity of life forms easily.
➢ The major characteristics taken for classify organisms are :
  1. Whether they are made up of prokaryotic or eukaryotic cells.
  2. They are made up of single or multi cells.
  3. Whether they made their food by own or not.
➢ The main credit of classification of organisms goes to Robert Whittaker.
➢ The organisms are classified into 5 kingdoms.

  1.Monera

  2.Protista

  3.Fungi

  4.Plantae

  5. Animalia

➢ The organisms are classified into five kingdoms:

  1. Thallophyta

  2. Bryophyta

  3. Pteridophyta

  4. Gymnosperm

  5. Angiosperm

➢ The kingdom anamilia are divided into 10 groups:

  1. Porifera

  2. Coelentreta

  3. Platyhelminthes

4. Nematode

5. Annelida

6. Arthopoda

7. Mollusca

8. Echinodermata

9. Protochordata

10. Vertebreta

➢ Vertebrata are divided into five sub classes:

1. Pisces

2. Amphibia

3. Reptelia

4. Aves

5. Mamallia

➢ Carolus linnaeus introduced the system of scientific nomenclature in the 18th century.

3. Diversity in living organisms

1. Who wrote the 'THE ORIGIN OF SPECIES'?

    (a)Robert hooke   (b)Robert frost

    (c)Robert Charles Darwin   (d)Robert brown

2. Which we called the region of mega diversity?

    (a) Tropic of cancer   (b) Tropic of capricon

    (c)Both(a)&(b)   (d)None of these

3. Who classify the organism into 5 kingdoms ?

    (a)Robert whittaker   (b)Carl woese

    (c)Frust haekel (d)Robert   (d) darwin

4. According to the whittaker the 4th kingdom is?

    (a)Animalia   (b)Plantae   (c)Fungi   (d)Protista

5. Mycoplasma is an example of which kingdom ?

    (a)Protista   (b)Monera   (c)Animalia   (d)Fungi

6. Cyarobacteria is known as ?

    (a)Mycoplasma   (b)Euglera

    (c)Blue-green algae   (d)Ipomea

7. Diatoms is an example of which kingdom ?

    (a)Monera   (b)Fungi

    (c)Plantae   (d)Protista

8. Fungi have cell walls made of tough complex sugar called ?

    (a)Chitin   (b)Cytosome

(c)Crystals   (d)Cycas

9.   Plants belongs to which group called algae ?

    (a)Bryophyta   (b)Petridophyta

    (c)Thallophyta   (d)None of these

10.   Which group is called the amphibians of the plant kingdoms ?

    (a)Thallophyta   (b)Bryophyta   (c)Petridophyta   (d)None of these

11.   Which of the following organisms cannot have cell wall ?

    (a)Neem (b)Spirogyra (c)Deodar (d)Hydra

12.   The animals have holes or pores all our the body are known ?

    (a)Coelantrata   (b)Phatyhelmenthes   (c)Arthopoda   (d)Porifera

13.   In porifera which helps in circulating water through out the body ?

    (a)Chitn   (b)Holes   (c)Pores   (d)(b)&(c)are same

14.   Spongilla are mainly found in ?

    (a)Terestrial habitat   (b)Marine habitat

    (c)Both (a)&(b)   (d)None of these

15.   Which of the following animal body made up of two ?

    (a)Porifera   (b)Sycon   (c)Planarians   (d)Hydra

16.   Which of these animal is covered with the a hard outside skeleton ?

    (a)Sycon   (b)spongilla (c)Hydra   (d)both (a) & (b)

17.   Which of the these group of animalia a pseudocoeln in present ?

    (a)Porifera   (b)Nematoda

    (c)Platyhelminthesis   (d)Coelentreta

18. Leeches and earthworms are example of ?

    (a)Arthopoda   (b)Annelida   (c)Nematoda   (d)Coelentreta

19. The largest group of animals ?

    (a)Arthopoda   (b)Nematoda   (c)Annelida   (d)None of these

20. Which of these have jointed legs?

    (a)Musca   (b) Butterfly   (c) None of these (d)Both (a) & (b)

21. Palaemon Arena and musca are the example of?

    (a)Annelida   (b)Nematoda   (c)Arthopoda   (d)None of these

22. Pariplanta is the scientific name of?

    (a)Housefly   (b)Butterfly   (c)Musca   (d)Cockroach

23. Aranea is the scientific name of?

    (a)Scorpion   (b)Spider   (c)Musca   (d)Cockroach

24. Asterias is the scientific name of?

    (a)Cockroach   (b)Butterfly   (c)Sea wichins   (d)Starfish

25. Starfish and sea urchins are the example of?

    (a) Arthopoda   (b)Echinodermata (c)Mollusca   (d)Nematoda

26. Which of the following is not the features of vertebrate?

    (a) Have a notochord   (b) Have a dorsal nerve cell

    (c) Have a gill bouched   (d) Are tripoblastic

27. Vertebrate are divides into _____ classes

    (a) 6   (b)5   (c)7   (d)4

28. Which of the following have only two chambered heart ?

(a)Petrio volitans   (b)Hyle   (c)Rana tigriana   (d)All of above

29.   Hyla belongs to which class ?

(a)Pisces   (b)Reptilia   (c)Amphibia   (d)All of above

30.   Scientific name of frog is?

(a)Hyla   (b)Rana   (c)Salamander   (d)Rana tigriana

31.   Amphibians are breath through?

(a)Lungs   (b) Gills   (c) Liver   (d) Both (a) & (b)

32.   Crocodile have _____chambers heart

(a)3   (b)4   (c)2   (d)None of these

33.   Crocodiles are belong to which group ?

(a)Reptilia   (b) vertebrate   (c)Arthopoda   (d)Plantae

34.   Which of these have mammary gland ?

(a)Whale   (b) Bat   (c) Human   (d)All of above

35.   Moss and Marelantra are the example of?

(a)Thallophyta   (b)Bryophyta

(c)Petridophyta   (d)Phanerograms

36.   Which of these have specialised tissue for conduction of food and water?

(a)Funaria   (b)Riccia   (c)Fern   (d)Chara

37.   Ferns ,Marsila and horse tails are the example of ?

(a)Arthopoda   (b)Mollusca   (c)Petridophyta   (d)Bryophyta

38.   Pires and deodar are the example of?

(a)Angiosperm   (b)Phanerogams

(c)Teridophyta   (d)Thallophyta

39. Hemidactylus is the scientific name of?

(a)Sparrow   (b)Lizard   (c)Pigeon   (d)Housefly

40. The scientific name of Flying lizard is ?

(a)Peluin   (b)Draco   (c)Both (a) & (b)   (d)Hyla

41. Which of the following is heterotrobh ?

(a)Musca (b)Ipomea (c)Moss (d)Ulothrix

42. Anabaena is the example of?

(a)Animalia   (b)Plantae   (c)Protista   (d)Monera

43. Scientific name of human is?

(a)Homo   (b) sapens   (c) Homospiens   (d) All of above

44. The system of scientific name is introduce?

(a)Charles Darwin   (b) Carolous linneus

(c) Both (a) & (b)   (d) None of these

45. The scientific name of white stork is ?

(a) Ciconia   (b)Ciconia-Ciconia

(c) None of these   (d) Camel

46. The scientific name of dog fish is ?

(a)Scoliodon   (b)String ray   (c)Torpedo   (d)None of these

47. Chameleon have _____ chamber heart
(a) 2   (b) 3   (c) 4   (d) None of these

48. Male hippocampus is the scientific name of?

(a) Flying fish   (b) Sea horse   (c) String ray   (d) Dogfish

49.   Blanoglassus is the example of?

(a)Nematoda   (b)Arthopoda   (c) Protochordata   (d) Annelida

50.   Herdmania and Amphioxus is the example of?

(a) Nematode   (b) Arthropod   (c)Protochordata   (d) Annelida

51.   Liver fluke belongs to which group?

(a)Nematoda   (b) Platyhelmenthes

(c) Annelida   (d) Arthopoda

ANSWERS:

Q	A	Q	A	Q	A
1	C	18	B	35	B
2	C	19	A	36	C
3	A	20	D	37	C
4	B	21	C	38	B
5	B	22	D	39	B
6	C	23	B	40	B
7	D	24	D	41	A
8	A	25	B	42	D
9	C	26	C	43	C
10	B	27	B	44	B
11	D	28	A	45	B
12	D	29	C	46	A
13	D	30	D	47	B
14	B	31	D	48	B
15	D	32	B	49	C
16	D	33	B	50	C
17	B	34	D	51	B

4. WHY DO WE FALL ILL

SOME IMPORTANT POINTS

- Health is a state of physical, mental, and social well being.
- The health of an individual is dependent on his/her physical surrounding and economic status.
- Diseases are characterised as a acute and chronic disease on their duration.
- Disease Kala-azar is caused by Leis mania.
- Viral disease is transmitted by contact by form a sick person to healthy person.
- Infectious agents are spread through air, water or vectors.
- Prevention of disease is best its successful treatment.
- Infectious disease can also be prevented by using immunisation.
- The category to which a disease-causing organism belongs decides the types of treatment.

4. WHY DO WE FALL ILL

1. The musculoskeletal system which is made up of.

   (a) Bones (b) muscles (c) bone and muscles   (d) none of these

2. Musculoskeletal system holds the body part together and helps the body

   (a)move   (b)stop   (c)rest   (d)none of these

3. A state of being well enough to function well physically.

   (a)health   (b)disease (c)none of these

4. Kidney is filtering.

   (a) beating   (b)fluring   (c)wine   (d)thinking

5. Energy and raw material are needed from --------------- the body.

    (a)inside (b)outside   (c)none of these

6. What is a necessity for cell and tissue functions?

    (a) Sun light   (b) air   (c) food   (d) oxidation

7. The environment includes the.

   (a)chemical environment   (b)physical environment

   (c)biological environment   (d)none of these

8. Health is at risk in a cyclone in ------------- ways.

   (a) short   (b)small   (c)many   (d)big

9. Our physical environment is decided by our.

   (a) Physical environment

   (b) Chemical environment

(c) Social environment

(d) biological environment

10. Food will have to be earned by doing.

(a) fighting   (b)energy   (c)work

11. Some diseases last for only very short periods of time called .

(a)chteronic diseases   (b)acute disease

(c)cuteronic diseases   (d)none of these.

12. Some diseases last for a long time even as much as a lifetime called.

(a)chronic disesases   (b)acute diseases

(c)cuteronic diseases   (d)none of these

13. Which is the example of chronic disease?

(a)Tuberculosis   (b)fever   (c)Aids   (d)a and c

14. Which is the example of chronic disease.

(a)fever   (b)T.B   (c) all diseases   (d)none of these

15. Loose motions is an example of

(a) acute diseases   (b)chronic diseases

(c) cuteronic diseases   (d)none of these

16. The disease that is non communicable is.

(a) Malaria   (b) marasmus   (c) AIDS   (d) jaundice

17. Malaria is caused by a?

(a)protozoan   (b)fungi   (c) virus   (d) bacteria

18. The hungi of the isabgol seed with water produces relief from?

(a)malaria   (b) flue   (c)diarrhea

19. Oral rehydration solution does not contain?

(a) sodium   (b)sodium bicarbonate

(c) Sodium cyanide   (d)glucose

20. The vitamin that is not fat soluble is ?

(a)vitamin A   (b)vitamin B complex

(c)vitamin D   (d)vitamin E

21. Xerophthalmia is caused due to the deficiency of?

(a)vitamin A   (b)vitamin B   (c)vitaminE

22. The 4D-syndrome characterizes the following disease?

(a)pellagra   (b)scurvy   (c)beriberi (d)xeropthalima

23. Maize interferes with the absorption of?

(a)ascorbic   (b)nicpotinic acid

(c)thiamine   (d)iodine

24. Sun light enhances the production of?

(a)vitamine A   (b)vitamine B

(c)vitamine C   (d)vitamine D

25. The proposed two in one salt condium iodine and

(a)sodium   (b)potassium   (c)iron   (d)manganese

26. An insect which transmits a disease is known as?

(a)intermediate host   (b)parasite   (c)vector   (d)prey

27. A chronic case of a diseases denotes?

(a) Severe attack of a disease

(b) Mild attack of the diseases

(c) Disease occurs for a very long period

(d) all of these

28. Which one of the diseases is not communicable?

(a)Typhoid   (b)diabetes   (c)hypertension   (d)helminthes

29. Which one of the diseases is none communicable?

(a)typhoid   (b)leprosy   (c)measles   (d)leukemia

30. Congenital diseases are those which?

(a) Are deficiency diseases

(b) Are present from time of birth

(c) Are spread from man to man

(d) Occur during life time

31. BCG vaccine is used to curb?

(a)Pneumonia   (b) Tuberculosis   (c)Polio   (d) Amoebiasis

32. AIDS virus has?

(a)Single strand DNA   (b) double strand DNA

(c)Single strand RNA   (d) double strand RNA

33. Causative agent of T.B.is?

(a)defective liver   (b)defective thymus

(c)AIDS virus   (d) weak immune system

34. Immuno deficiency syndrome could develop due to ?

(a) Defective liver   (b) defective thymus

(c)AIDS virus   (d)weak immune system

35. T.B. is cured by?

(a) grise ofulvin   (b) ubiquinone   (c) encitol   (d) streptomycin

36. AIDS is due to?

(a) Reduction in number of helper t- cell

(b) Reduction in number killer   t-cell

(c) Auto immunity

(d) non production of intererons

37. Typhoid is caused by?

(a)Escherichia   (b)giardia   (c)salmonella   (d)shigella

38. Which of the following is a mismatch?

(a) Leprosy – bacterial infection   (b) AIDS–bacterial infection

(c) Malaria – protozoan infection   (d) elephantiasis- nematode infection

39. Calcium deficiency occurs on the absence of vitamin?

(a) D   (b)A   (c)C   (d)B

40. Fever, delirium, slow pulse, abdominal tenderness and rose colored rash Indicate the disease?

(a)typhoid   (b) measless   (c)tetanus   (d)chicken pox

Answers:

QUES.	ANS.	QUES.	ANS.	QUSE.	ANS.	QUES.	ANS.
1	C	11	B	21	A	31	B
2	A	12	A	22	A	32	C
3	A	13	D	23	B	33	B
4	C	14	A	24	D	34	C
5	A	15	A	25	C	35	D
6	C	16	B	26	C	36	A
7	B	17	A	27	C	37	C
8	C	18	C	28	A	38	B
9	C	19	C	29	D	39	A
10	C	10	B	30	b	40	A

NOTES

www.ingramcontent.com/pod-product-compliance
Lightning Source LLC
Chambersburg PA
CBHW070744180526
45168CB00004B/1521